P9-AQZ-905

earth issues

THE ENERGY DILEMMA

BY JESSICA GUNDERSON

CREATIVE EDUCATION

Published by Creative Education
P.O. Box 227, Mankato, Minnesota 56002
Creative Education is an imprint of
The Creative Company
www.thecreativecompany.us

Design and production by The Design Lab
Art direction by Rita Marshall
Printed by Corporate Graphics in the
United States of America

Photographs by Alamy (Ashley Cooper), Dreamstime (Owen B. Lever, Olexander Pelypenko, Pniesen, Aschwin Prein, Yevgen Stoletnij, Richard Wilson), Getty Images (Sisse Brimberg, Brian Lawrence, STR/AFP), iStockphoto (Rob Belknap, Rob Broek, Mark Hayes, Eric Hood, Justin Horrocks, Rafa Irusta, Björn Kindler, Maciej Noskowski, Andy Olsen, Jim Parkin, James Richey, Todd Smith, Kathy Steen, Andrey Volodin, Nicole Waring)

Library of Congress
Cataloging-in-Publication Data
Gunderson, Jessica.
The energy dilemma / by Jessica Gunderson.
p. cm. — (Earth issues)
Includes bibliographical references and index.
Summary: An examination of the resources humans use to create energy, exploring fossil fuels' impacts on the environment and discussing cleaner, more sustainable options that may contribute to a healthier planet.
ISBN 978-1-58341-980-9
1. Power resources—Juvenile literature. 2. Power resources—Environmental aspects—Juvenile literature. 3. Pollution prevention—Juvenile literature. I. Title. II. Series.

TJ163.23.G86 2010
333.79—dc22 2009028048

CPSIA: 120109 PO1091
First Edition
9 8 7 6 5 4 3 2 1

Table of Contents

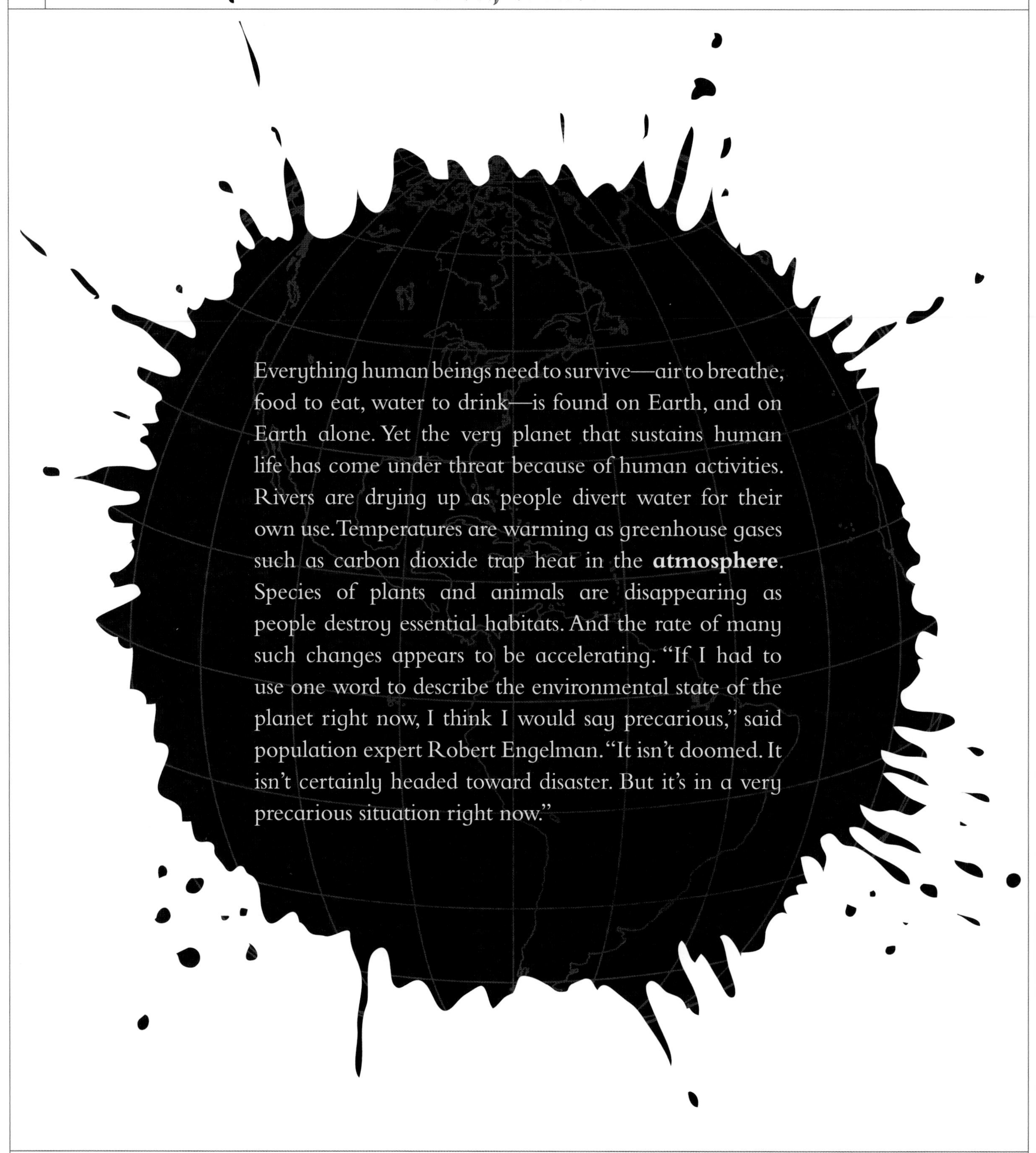

Everything human beings need to survive—air to breathe, food to eat, water to drink—is found on Earth, and on Earth alone. Yet the very planet that sustains human life has come under threat because of human activities. Rivers are drying up as people divert water for their own use. Temperatures are warming as greenhouse gases such as carbon dioxide trap heat in the **atmosphere**. Species of plants and animals are disappearing as people destroy essential habitats. And the rate of many such changes appears to be accelerating. "If I had to use one word to describe the environmental state of the planet right now, I think I would say precarious," said population expert Robert Engelman. "It isn't doomed. It isn't certainly headed toward disaster. But it's in a very precarious situation right now."

Human beings' hunger for energy is only worsening the threats to the environment. We require energy for nearly everything we do, and much of this energy comes from fossil fuels such as coal, oil, and natural gas. Mining fossil fuels destroys natural landscapes and habitats, and burning fossil fuels releases greenhouse gases into the air. As the supply of fossil fuels runs low, the world is looking for more **sustainable** and environmentally friendly sources of energy. But in the meantime, how do we curb our appetite for energy? Are fossil fuels as harmful as some experts believe? How do we transition from fossil fuels to clean energy sources, and what will be the cost?

Energy is the driving force behind all human life. We use energy stored in our bodies to move, breathe, and eat. We also use energy from the earth to heat our homes, power our appliances, and run our vehicles. Much of that energy comes from fossil fuels such as coal, oil, and natural gas, which are found beneath the earth's crust. Fossil fuels are nonrenewable sources of energy, which means that they will run out someday. At the rate at which humans are using fossil fuels, scientists estimate that the world's supply could run out by the year 2125.

CHAPTER ONE

An Appetite for Energy

Although fossil fuels are being depleted quickly, they took a long time to form. They were created millions of years ago from the remains of plants and animals. As time passed, the decayed remains became buried under layers of sediment and other **organic** matter. The layers pressed tightly together, causing high-pressure and high-temperature conditions and turning the remains into fossils. Different types of fossil fuels were formed depending upon the type of organic matter, pressure level, and temperature involved. Coal was formed from decayed organic remains on land, while oil and natural gas were formed from decayed marine life below the prehistoric seas.

Coal, natural gas, and oil contain an element called carbon, which makes them flammable. People burn them in order to generate electricity or to power machinery and automobiles. More than three-fourths of the world's energy comes from fossil fuels. Coal-burning power plants generate electricity for home and commercial uses. Natural gas is used mostly for heating and cooking purposes. Oil is **refined** to make gasoline for automobiles and airplanes.

Testing the substances emitted by cars in urban areas such as Atlanta, Georgia, has become necessary to help curb pollution in recent years.

After the Industrial Revolution took hold in America during the late 1800s, many people worked in fossil-fuel-powered factories.

Before the mid-18th century, people used only small amounts of fossil fuels. Early civilizations often used wood for heating and cooking; after coal was found to be flammable, people began mining the material and using coal in stoves and furnaces. Kerosene, a fuel made from oil, was used in lamps. During the 17th and 18th centuries, people also relied on other sources of energy, such as wind and water. Windmills pumped water from the ground for **irrigation** purposes. Cotton and textile mills in Great Britain and elsewhere relied on the power of rushing water to turn large wheels that were connected to machinery inside the factories.

The **Industrial Revolution**, however, accelerated people's use of fossil fuels. Inventors developed machines that could manufacture goods, transport people, and help perform household activities. These machines were often powered by fossil fuels. The steam engine, developed for widespread use in 1765, was one of the major contributors to the Industrial Revolution. Fueled by coal, the steam engine provided the means to power large machinery such as trains and textile **looms**. By the 1800s, electricity was being developed for widespread industrial use. The first power plants to distribute electricity ran on coal.

Steam engines were the most commonly used power sources for locomotives in the U.S. from the 1830s until nearly a century later.

Fueling the Fire

The World Energy Council (WEC), established in 1923, has members in almost 100 countries. The WEC tackles issues regarding all types of energy—coal, oil, natural gas, **nuclear**, and renewables—and is the leading energy organization in the world today. Its mission is "to promote the sustainable supply and use of energy for the greatest benefit of all people." To do so, the WEC provides reports, studies, energy-use projections, and policy recommendations to world governments, with the goal of encouraging officials to adopt energy plans that will help people reach sustainable solutions that cause the least amount of damage to the earth.

Early cars underwent dramatic changes after Karl Benz's 1885 model, thanks largely to the assembly-line process instituted by Henry Ford.

With the advancements of the Industrial Revolution came an increased demand for fossil fuels. By the 20th century, people in many parts of the world were using fossil fuels in everyday life. Automobiles, invented in 1885, and airplanes, invented in 1903, use an oil-based fuel called gasoline. These vehicles eventually made travel faster and more accessible, enabling people to make more trips and get around with greater ease. By the 1950s, the demand for oil had grown so high that it was being pumped from every continent except Antarctica. Technology improved in the 20th century to keep pace with demand and increased people's use of electricity in the home. People could spend their time using computers, watching television, or listening to radios and stereos, all of which use electricity.

Surface mining methods such as open-pit mining require access to vast stretches of coal—and many dump trucks to transport it.

Today, there are more than 50,000 coal-fired power plants in the world. Most of the world's coal supplies are located deep underground. Machines burrow a tunnel to reach the coal supply, or seam. Miners then ride elevators down the tunnel and use machines and other equipment to extract the coal. Sometimes coal is found closer to the surface of the ground, and coal miners use modern surface mining techniques such as strip mining or area mining. In surface mining, the ground covering the coal is removed by machines, and the coal is extracted without the use of tunnels. The coal is then transported to power plants, where it is burned to heat large vats of boiling water. The steam from the boiling water spins **turbines** that are connected to generators. The generators produce electricity, which is supplied to homes, businesses, and factories through power lines.

Fueling the Fire

In March 1989, the *Exxon Valdez* oil tanker ran aground in Prince William Sound, Alaska, spilling 11 million gallons (41.6 million l) of oil. The spill was not the largest in history, but it is widely believed to have been the most environmentally damaging due to the remote location, which made it difficult for cleanup workers to act quickly. More than 250,000 seabirds, 2,800 sea otters, and 300 harbor seals died from the oil's harmful effects. Hundreds of bald eagles and 22 killer whales also perished. Today, local populations of the Pacific herring and the pigeon guillemot have yet to recover.

When an oil spill occurs in a body of water such as a harbor, the oil slick spreads across the surface, polluting everything in its path.

Oil is extracted from wells that are formed by drilling into the ground. Well pumps push the oil into pipelines, which carry it either directly to refineries—where it is changed into substances such as gasoline, asphalt, and lubricants—or to ships, trucks, and trains, which then carry it to refineries. Natural gas, often found with oil underground, is also extracted from wells. Natural gas can be used in its pre-existing form and is piped out for use in furnaces, water heaters, and stoves.

Fossil fuels will not be available forever, but this is not the only problem with depending upon them as our primary energy source. Burning fossil fuels creates pollution and contributes to global warming. When burned, fossil fuels release carbon dioxide, a greenhouse gas, into the atmosphere. Greenhouse gases act as a blanket around Earth, keeping the planet at a temperature that can sustain life. But when the concentration of greenhouse gases becomes too high, it can cause the planet to gradually warm. Before the Industrial Revolution, there were 280 parts per million (ppm) of carbon dioxide in the air, but now the amount of carbon dioxide has risen to 380 ppm. As more fossil fuels are burned, more carbon dioxide is released, which contributes to a warmer planet. Earth's average temperature has risen 1 °F (0.6 °C) over the past century, and experts expect it to continue rising. That number may seem insignificant, but even a small increase in overall temperature can have a large impact on the planet, causing catastrophic storms, droughts, and flooding.

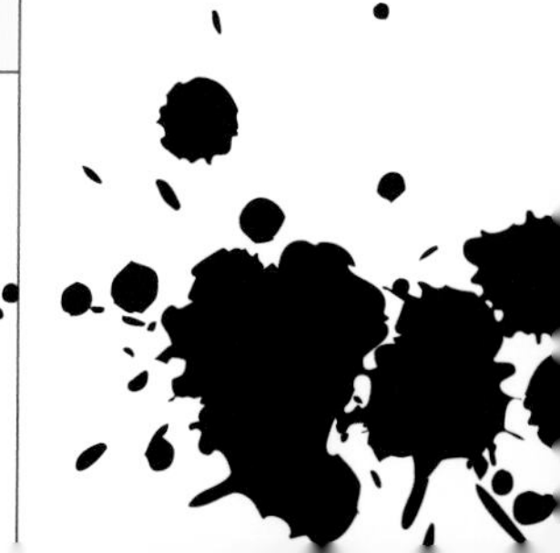

Today, the United States uses more energy than any other country in the world, and almost all of that energy is produced by fossil fuels. In 2008, global energy consumption increased by 1.4 percent from 2007 levels, with the U.S. consuming 20.4 percent of the world's total energy. China's use of energy climbed 7.2 percent from 2007 levels. However, North America, including the U.S., decreased its energy consumption during the same time period by 2 percent. In the U.S. alone, consumption dropped 2.8 percent, the largest decline since 1982.

CHAPTER TWO

Fossil Fuel Dependence

By the year 2030, world energy consumption is expected to expand by 50 percent. This anticipated increase is attributed to economic growth and rising populations of the world's **developing countries**, such as China and India. As developing countries become developed, or **industrialized**, their **economies** grow, which usually means citizens are able to make more money. When their standard of living improves, people tend to buy their own homes and cars, which in turn increases their energy consumption.

Oil is a primary player in the energy game. Global demand for oil reached 84.5 million barrels per day in 2008, 0.6 percent less than in 2007. The U.S. consumed almost one-fourth of this amount, at about 19.4 million barrels daily. The decrease in global consumption, the first since 1993, was chiefly due to a widespread economic downturn and increased awareness of the costs of energy use. If this trend continues in future years, the world may see less reliance upon oil and a decline in oil prices. In 2008, the price of **crude oil** averaged $100 per barrel, which was a 37.7 percent increase from 2007 prices. Between 2002 and 2009, the price of oil in the U.S. nearly tripled. The price of oil

Lacking in most natural resources, Hong Kong became a center of trade and finance by necessity, with a highly industrialized economy.

dropped sharply in the second half of 2008, however, falling to $40 per barrel.

Variation in oil prices can be attributed to many factors. One factor involves the law of supply and demand. When the demand for oil is higher than the supply, prices increase, but when the demand decreases, so do prices. In the latter part of 2008, supply exceeded demand, causing the price of crude oil to decline, though the average price of oil for the whole of 2008 was still higher than that of 2007. Natural disasters can also affect oil prices because storms such as 2005's Hurricane Katrina threaten oil reserves, many of which are located offshore. Wars in oil-rich countries, such as the Persian Gulf War in the early 1990s, create instability in the oil market, leading to higher oil prices.

Although oil was formed under the sea millions of years ago, many oil supplies are now located beneath land due to receding oceans and movements of the earth's crust. Oil can be found in many countries, but some countries are more oil-rich than others. Most nations have to import oil to meet their energy needs, causing them to depend heavily upon those countries that supply the fuel. The 12 oil-rich countries of the Organization of Petroleum Exporting Countries (OPEC), a body that was formed in 1960, regulate oil prices and policies, controlling and protecting the business of oil exportation. Because OPEC exercises such power, the organization's actions are often open to debate. Many nations, such as the U.S., are trying to gain energy independence rather than continuing to rely on oil imports.

Offshore oil rigs in the ocean are situated on sprawling platforms that have to house workers as well as machinery for drilling.

Fueling the Fire

Because burning coal produces large amounts of carbon dioxide, chemists are continually looking for ways to make coal cleaner. One innovation in clean coal technology is called carbon capture and storage. The carbon dioxide emitted from the burned coal is captured, usually by means of chemicals that absorb the gas, before it can be released into the air. The carbon dioxide is then compressed into a liquid and piped to an underground storage unit, never to be released into the atmosphere. Scientists are also looking for ways to increase coal-fired power plants' efficiency, which will lower overall carbon dioxide emissions.

Coal is often stockpiled in great quantities near a power plant in order to provide a steady and easily accessible supply of fuel.

Coal is more abundant than oil and can be found in many parts of the world, and because it is inexpensive and widely available, consumption of coal-based energy is steadily climbing. Coal provides 28 percent of the world's total energy needs and generates 40 percent of the world's electricity. Coal is also an important ingredient in the production of steel. About 13 percent of the world's hard coal is used in the steel industry. China is the world's largest producer and consumer of coal, using 42.6 percent of the world's total supply in 2008. The U.S. uses the second-highest percentage of coal, at 17.1 percent.

While coal is the most abundant fossil fuel, it is also the dirtiest. The burning of coal releases a toxic gas called sulfur dioxide. Sulfur dioxide mixes with water in the air and causes the precipitation that falls to Earth to turn acidic. In the late 1970s and early '80s, scientists and fishers noticed a decline in animal and plant life in parts of the eastern U.S. In 1981, the National Academy of Sciences released a report that attributed that decline to acid precipitation, much of which was produced by coal-fired power plants. Acid precipitation removes nutrients from the soil and damages leaves, limiting trees' and plants' nutrient intake and making them vulnerable to diseases. Acid precipitation falling into lakes and rivers makes the water more acidic, and acidic water carries a low amount of oxygen. When fish and other marine animals cannot get enough oxygen, they suffocate and die. The U.S. Clean Air Act of 1990 called for power plants to reduce their sulfur dioxide emissions by 50 percent within 10 years. This act was a significant step in reducing acid precipitation pollution in the U.S., but other coal-burning countries such as China have few policies that limit emissions from their coal-burning power plants and factories, so the pollution continues to travel around the world.

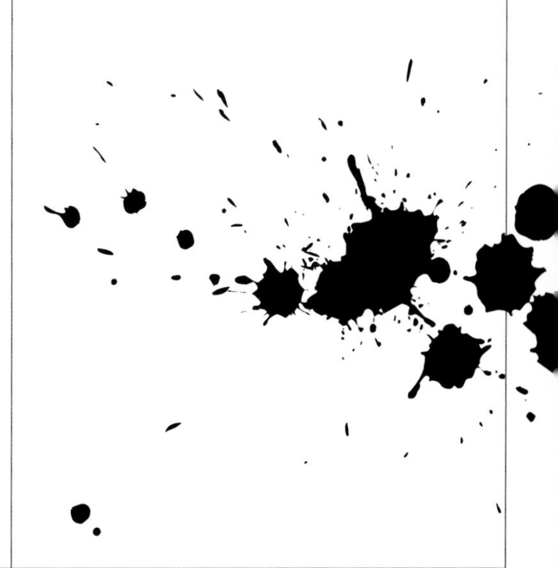

In 2009, Great Britain announced that any new coal-fired power plants would be required to capture and store carbon emissions.

Another problem with coal-burning power plants is that they are not very efficient. At least 50 percent of the energy contained within the coal is lost when it is burned. In the U.S., the average power plant operates at only 33 percent efficiency, which means that it loses 67 percent of the energy contained in the coal. When coal is burned to boil water, it produces steam to spin the turbines, but some of the heat escapes to warm the air instead and is released through boiler chimneys. The steam that is used to spin the turbines eventually cools and is transferred to cooling towers, where it cools further and condenses into water. The warm water is then emptied into a nearby river or lake. Still more energy is lost during coal's **conversion** from mechanical energy (the spinning of the turbines) to electrical energy. By the time the electricity reaches people's homes, about two-thirds of the coal's energy has been lost.

The carbon dioxide emitted from burning fossil fuels is of major concern to scientists and others around the world. Evidence points to carbon dioxide as one of the major causes of global warming. Global warming has been linked to various disastrous effects, such as major melting in the Earth's polar regions. Polar animal species, such as polar bears and penguins, that rely on cold temperatures and ice and snow in order to live are being threatened by the melting of ice sheets and subsequent rising sea levels. In the north, the ice shelves and floes that polar bears depend upon to reach their prey have been melting, and the gradually warming weather has caused ice to form later in the year and melt earlier, disrupting the bears' feeding habits and causing some to die of starvation. Near the South Pole, melting snow and ice can trap penguins and keep them from moving to their feeding and breeding grounds when the seasons change.

CHAPTER THREE

Energy and the Environment

Even without taking the melting ice into consideration, sea levels are already rising. Water expands as it warms, so warmer water covers more area than cold water, causing sea levels to rise. Warmer water also threatens many marine species that are not designed to live in warm water, so they must move to colder waters or face **extinction**. Warming oceans also affect weather patterns, causing more strong storms and hurricanes to develop. Major storms can destroy homes, lives, and habitats. In 2005, Hurricane Katrina swept through the Gulf of Mexico, killing hundreds of people and displacing thousands more who lived along the coastlines. Some believe that global warming was one of the causes behind the hurricane.

The actions of drilling and mining for fossil fuels also have negative effects on people and the environment. Underground

Polar bears hunt seals from the vantage point of an ice floe, often waiting several hours for an animal to appear at the surface.

Fueling the Fire

The Kyoto Protocol is an international treaty that calls for industrialized countries to reduce greenhouse gas emissions by an average of 5.2 percent below 1990 levels by 2012. The **ratification** of the Protocol by 184 countries—with the notable exception of the U.S.—has led to a near-global attempt at reducing fossil-fuel-based energy. One way a country can meet the demands of the Protocol is through emissions trading. Countries with emissions levels below the target percentage earn credits. They may then sell those credits to countries whose emissions are too high. This practice benefits countries with low emissions and the planet overall.

The 2005 Sunjiawan mine disaster was 1 of more than 3,300 explosions, fires, or floods that occurred in Chinese mines that year.

coal mining is dangerous to the miners who must travel underground to extract the coal, as they are forced to inhale harmful gases and face the possibility of cave-ins and explosions. Many coal miners develop black lung disease from breathing coal dust, while accidents have claimed the lives of others. In 2005, an explosion at the Sunjiawan mine in China killed more than 200 workers.

Surface mining techniques used to extract lignite coal (a low-grade variety of brown coal) often displace large amounts of land, disturbing both wildlife habitats and fields used for agriculture. Even though coal companies often replace the soil after the coal has been mined, it can take years for the land to return to its natural state. Mining produces large amounts of waste, which is often dumped into piles. Precipitation runs off these piles and can carry pollutants into area lakes and rivers.

With oil, most of the environmental damage can occur when the product is transported. Oil must be taken great distances to reach refineries; it crosses land on trucks, trains, and pipelines, and it is carried across oceans on large ships called oil

More than 15 billion barrels of oil have been transported via the Trans-Alaska Pipeline System since it was put into use in 1977.

tankers. While in transit, accidents may occur that cause oil spills. One of the largest oil spills in history occurred in 1979 when two oil tankers collided in the Caribbean Sea, spilling about 316,363 tons (287,000 t) of oil and damaging marine life in the area. Sea birds coated in oil cannot fly, and they sometimes ingest the oil and die. Fish and other creatures in the sea are likewise poisoned by the oil. Oil spills also negatively affect the economies of nearby countries, especially those that rely on fishing and tourism.

Oil is often called "black gold" because it is a precious resource that can bring riches to those who possess it. With oil comes power and wealth, and many people want the power that oil-rich countries have. In August 1990, Iraq invaded the oil-rich country of Kuwait, with hopes of controlling its oil and acquiring its wealth. Other countries, such as the U.S. and Great Britain, concerned about their own access to Kuwait's oil, went to war against Iraq in January 1991. Kuwait was saved, but thousands of lives were lost. Many Kuwaiti oil wells were damaged due to bombing in the region. After the war ended, oil fires burned for many months.

Other conflicts have been waged over oil around the world. In the African country of Nigeria, violent clashes have occurred between **ethnic** groups over control of the oil-abundant region around the Niger River Delta. Conflicts over oil have also arisen in the Caspian Sea republics of Kazakhstan and Chechnya, and in Southeast Asia and South America. As the demand for oil grows higher, more and more countries are fighting for control of the valuable resource.

The oil fires that continued to burn in Kuwait after the Persian Gulf War ended in 1991 contributed to pollution in the atmosphere.

The concrete cooling towers at a nuclear power plant remove excess heat from the circulating water used during the fission process.

Nuclear energy, which is produced by a process called nuclear fission, has been suggested as an alternative to fossil fuels because it is cleaner and more efficient. During nuclear fission, atoms of a metal called uranium are split, and small particles called neutrons are released. These neutrons bump against other atoms, splitting them and resulting in a chain reaction of nuclear fission. Fission releases heat that is used for boiling water and producing the steam that spins turbines and generates electricity. About 18 ounces (500 g) of uranium produces the same amount of heat as 1,400 tons (1,270 t) of coal. Currently, nuclear energy powers about 14 percent of the world's electricity. France, which receives about three-fourths of its power from nuclear sources, is the leader in nuclear energy consumption. Nuclear power plants release a significantly lower amount of the heat-trapping gas carbon dioxide than fossil fuel power plants do.

However, nuclear energy is not without its problems. One of the main controversies about nuclear energy revolves around the **radioactive** waste the power plants produce. Some waste is classified as low-level and consists of trash such as clothing, rags, and tools that have been contaminated by coming into contact with radioactive materials. Other waste is high-level and made up mostly of spent nuclear fuel, the radioactive remains of the fuel used by

Areas such as this abandoned playground in Chernobyl were contaminated by the radioactivity in the air after the nuclear disaster.

Fueling the Fire

An accident at the Chernobyl Nuclear Power Plant in Ukraine was the worst nuclear disaster in history. In April 1986, due to a testing mistake, one of the plant's reactors exploded, causing a chain of several other explosions, which emitted radioactive particles into the air. The radioactive cloud spread across parts of Russia (then known as the Soviet Union) and eastern Europe. More than 350,000 people were evacuated, including the entire city of Chernobyl. It is estimated that more than 125,000 people may have died from being exposed to radioactivity. The disaster raised concerns worldwide about the safety of nuclear energy and slowed its development.

The Chernobyl reactor that exploded was encased in a concrete-and-steel structure to prevent further emissions of radioactive particles.

nuclear power plants. Exposure to radioactive waste can cause cancer, birth defects, and reproductive problems in humans. The waste remains radioactive for thousands of years, and there is no way for radioactive waste to be eliminated, recycled, or otherwise disposed of. Most low-level waste is stored in special, leak-proof containers and buried deep underground, while high-level waste is kept temporarily (but indefinitely) at the power plants. Apart from the wasteful byproducts, use of nuclear energy also poses a security risk. Terrorists, or people who use violence and intimidation to get the things they want, could possibly sabotage nuclear power stations to set off an explosion, or they could steal radioactive materials to make nuclear weapons and explosives.

As the problems surrounding fossil fuels become larger, most experts believe we must wean ourselves off them and begin using clean, renewable energy sources, such as hydro (water), wind, solar, and geothermal. These sources are considered clean, or green, because they do not create as much pollution or emit carbon dioxide like fossil fuels do. Today, only about 1.5 percent of the world's electricity comes from such clean energy sources, but this rate is rising as the cost and consequences of using fossil fuels become higher and more apparent. Scientists, investors, and power companies are all looking toward renewable energy solutions in order to meet the world's growing energy demands.

CHAPTER FOUR

Fueling the Future

Hydropower is already a widely used source of energy, especially when it comes to generating electricity. Flowing water contains a great amount of **kinetic energy** that can be harnessed to produce electricity. Dams built on rivers store water in natural or man-made reservoirs. The water can be released through turbines to create electricity. Harnessing power from the ocean's tidal waves is another source of hydropower. Barriers called tidal barrages are built across **estuaries** and bays. When the tide comes in or out, water pushes through the barrages' turbines, producing electricity. Although hydropower is efficient and does not release greenhouse gases, it can damage the environment. Reservoirs created by dams flood large areas of land, destroying natural habitats. Dams change a river's natural course, flooding land and interfering with species that are native to the river. Tidal barrages often flood coastal marshes, which are feeding grounds for many seabirds, rendering the areas unusable.

The sun is a boundless source of energy that humans have only just begun to explore. Solar energy is available nearly everywhere on

The Colorado River's 726-foot-tall (221 m) Hoover Dam, completed in 1936, is the highest dam of its kind (a concrete arch construction) in the U.S.

Oceangoing ships and other craft can pass through the ship locks of the Three Gorges Dam on their way to inland cities and seaports.

Fueling the Fire

The Three Gorges Dam on China's Yangtze River is the largest hydroelectric dam in the world. The dam wall is 7,660 feet (2,335 m) long and creates a reservoir 375 miles (604 km) long. The dam provides electricity to nine Chinese provinces and represents about one percent of China's electricity generation. Although hydroelectricity is a renewable resource, the dam project was not without controversy. About 262 square miles (679 sq km) were flooded to make the dam, requiring more than 1.2 million people to relocate and disrupting hundreds of animal species that lived in the area's ecosystems.

Placing solar panel awnings on the sunniest side of a house can not only provide shade but also harness the most direct energy.

Earth, but it is most effective in areas that receive an abundance of steady sunshine. Solar panels, made up of a collection of solar cells that turn the sun's energy into electrical energy, can be placed on homes, power plants, and other buildings. Solar cells are made of thin layers of **silicon**. When sunlight hits the cell, the silicon absorbs the sun's energy, and electrical charges move between the silicon layers to produce an electric current. Some problems with solar energy are that panels are expensive to install and they require coverage over a large area in order to work efficiently. Even though the sun has enough energy to supply the world with electricity 10 times over, in 2008, only .05 percent of the power in the U.S. was solar-generated.

The wind is another abundant energy source. In recent years, wind energy farms have sprung up around the world. Wind spins the blades on giant

From enabling communications and navigation to providing weather images and military data, satellites are useful objects to have in space.

Fueling the Fire

By 2030, Japan hopes to launch an innovative idea to capture solar energy in space. Solar panels would be placed on a satellite that orbits Earth. The panels would collect the energy and transmit it, either by waves or lasers, to a receiving station on Earth. Advantages to using such a method include the fact that solar energy is stronger in space, and the satellite could continuously capture the sun's waves—unaffected by Earth's rotations or cloud cover. Japan is not the only country considering the idea; in 2008, the U.S. announced its plans to develop a test satellite that would beam solar energy to Earth.

Wind energy advocates hope to see the day when wind farms are widespread enough to generate 20 percent of America's electricity.

windmills, or wind turbines, turning the wind's kinetic energy into electrical energy. Windy areas of the world, such as the Great Plains region and the state of Texas in the U.S. and the Romanian province of Dobrogea, are home to the largest onshore wind farms. In Denmark, wind farms have even been built offshore, in the Baltic Sea. Wind power generates a little more than one percent of the world's energy, and that rate continues to increase. Yet although wind power may be a bountiful source of energy, it is not always a reliable one. When the wind dies down, the turbines stop spinning and no longer produce energy. To combat this problem, many wind turbines today are equipped with computerized motors that keep the turbines moving in the absence of wind.

Both natives and tourists enjoy the experience of bathing in the geothermal waters of southwestern Iceland's Blue Lagoon.

Heat from within the earth's core can be used to create geothermal energy. In many volcanic areas, heat lies closer to the earth's surface and often produces hot springs. Steam from the hot springs can be used to turn turbines to create electricity. Another method of retrieving heat is to inject cold water into holes bored deep in the ground. As the water heats, it produces steam. Much care must be taken when handling geothermal energy so as to avoid mixing the water used in the process with underground freshwater supplies. Another drawback to geothermal energy is that it is not easily accessible in all countries. Iceland, which has a high concentration of volcanic areas, is rich in geothermal potential, and geothermal power heats nearly 90 percent of the country's buildings.

Fueling the Fire

Iceland, an island country located in the North Atlantic Ocean, is the "greenest" country on the planet. Most of its electricity is generated from renewable geothermal and hydroelectric sources. Since Iceland has an abundance of volcanic areas, where heat from the earth's core lies close to the surface, residents have easy access to geothermal energy. Additionally, most homes are heated with water pumped from geothermal hot spots, while Iceland's rivers and waterfalls make it a prime candidate for hydroelectric power. Virtually the only fossil fuels used there are for powering automobiles and fishing boats, but the country has pledged to become fossil-fuel-free by 2050.

Geothermal power stations are a common sight throughout Iceland, due to the country's volcanic and geyser-studded landscape.

Many ethanol production plants in the Midwest are located near fields of corn—the most common basis of the fuel.

In addition to using clean energy for industrial and home use, most experts believe that another way to combat the energy dilemma and lessen our dependence on fossil fuels is to use alternative energy to power automobiles. Biofuels, or renewable fuel sources made from organic matter, are currently undergoing research and development. One biofuel already in widespread use is ethanol. Ethanol is made from crops such as sugar cane, sugar beets, and corn, as well as from organic wastes. Ethanol is a clean-burning fuel that is commonly blended with gasoline for use in flex-fuel automobiles, or vehicles equipped to burn both types of fuel.

Since the 1990s, engineers have been developing cars that will burn fuel more efficiently. Hybrid cars use fuel-powered engines at high speeds and electric engines at low speeds, reducing the need for gasoline and producing lower emissions. Fuel-cell cars run on battery power. The fuel-cell battery combines hydrogen and oxygen to make electricity, which powers the motor. The only wastes produced from a fuel cell are water and negligible amounts of carbon dioxide. As fuel-cell cars have become more popular, more hydrogen refueling stations have opened around the world, with most located in Iceland, Germany, and Japan in the first decade of the 21st century.

People around the world are excited about recent innovations in clean-energy technology; however, it will take a long time to make a complete shift away from fossil fuels. Much of the world will continue to rely on fossil fuels because they are cheaper and more accessible. In addition, industries centered on fossil fuels employ millions of people, and considering that so many countries' economies rely on the export, import, mining, and burning of fossil fuels, a global shift toward clean energy is highly controversial. Even though clean energy development will eventually create more jobs, those who are currently employed or invested in the fossil fuel industries fear losing their jobs or money.

Still, many fossil-fuel-based companies are investing in alternative energy in efforts to become more energy conscious and create more jobs. For example, since 2005, oil and natural gas company British Petroleum (BP) has invested $2.9 billion in alternative energy sources such as biofuels, wind, solar, and hydrogen. Many electric companies in the U.S. and Europe offer local customers the opportunity to buy power produced by green energy instead of fossil fuels, and the trend is widening.

As the world moves toward embracing clean, renewable energy sources, the energy dilemma of the last few decades could become a thing of the past. With the help of scientists, engineers, corporations, and individuals, Earth may once again become a place of boundless energy opportunities, sustaining all life for centuries to come.

In addition to being grown in natural settings, algae can also be cultivated in man-made systems such as this one by AlgaeLink.

Fueling the Fire

One of the most promising biofuels of the future is algae-based. Algae are fast-growing plants that often live on the surface of ponds and lakes. The plants are rich in oils, containing about 30 times more oil than sunflower seeds, and these oils can be refined into biofuel. Another benefit of using algae for fuel is that it can be grown on nearly any body of water and does not use up land that could be used for food crops instead. In January 2009, the first airplane to use a fuel made partly from algae oils took off from a Houston, Texas, airport.

Fueling the Fire

Changing energy prices and increasing environmental concerns are causing people to think more about how they use energy today, and many individuals are becoming more energy-conscious. Instead of driving everywhere, people can conserve energy by walking or biking whenever possible. Turning off lights, appliances, and computers when they are not in use is a good way to conserve electricity, which also lowers carbon dioxide emissions. And turning down the thermostat a few degrees in winter and raising it a few degrees in summer not only lowers energy usage, but it also helps save on heating and cooling costs.

People who can ride a bicycle to meet their daily transportation needs help the earth by producing neither pollution nor waste.

Glossary

atmosphere—the layer of gases that surrounds Earth

conversion—the process of changing from one form or function to another

crude oil—oil in its natural state, before it is made into gasoline and other products

developing countries—the poorest countries of the world, which are generally characterized by a lack of health care, nutrition, education, and industry; most developing countries are in Africa, Asia, and Latin America

economies—the social systems that involve production and consumption of countries' wealth, goods, and services

estuaries—the wide mouths of rivers where they join the sea

ethnic—relating to large groups of people with common racial, national, religious, or cultural backgrounds

extinction—the condition or process of a species or larger group no longer existing

Industrial Revolution—a period during the late 18th and early 19th centuries in Europe and the U.S., marked by a shift from economies based on agriculture and handicraft to ones dominated by mechanized production in factories

industrialized—having highly developed industries or manufacturing activities

irrigation—the distribution of water to land or crops to help plant growth

kinetic energy—energy produced by motion

looms—frames or machines for interlacing threads to form cloth

nuclear—having to do with energy created when atoms are split or joined together

organic—derived from or relating to living matter

radioactive—characteristic of substances such as uranium that give off particles of energy as their atoms decay; the energy is dangerous to human health

ratification—formal approval of a treaty or agreement so that it can take effect

refined—processed or changed; crude oil is refined into usable products such as gasoline and lubricants

silicon—a chemical element found in sand and rocks, often used in making electronics, glass, and cement

sustainable—using a resource in a way in which it is not depleted or permanently damaged

turbines—machines that are driven by water, steam, or a gas flowing through the blades of a wheel

Bibliography

Coal Utilization Research Council. "Clean Coal 101: What Is Clean Coal Technology?" www.coal.org/clean_coal_101/index.asp.

Darley, Julian. *High Noon for Natural Gas: The New Energy Crisis.* White River Junction, Vt.: Chelsea Green Publishing, 2004.

Elliot, David. *Energy, Society, and Environment*. 2nd ed. New York: Routledge, 2003.

Evans, Robert. *Fueling Our Future*. New York: Cambridge University Press, 2007.

Flannery, Tim. *The Weather Makers: How Man Is Changing the Climate and What It Means for Life on Earth*. New York: Atlantic Monthly Press, 2005.

Krupp, Fred, and Miriam Horn. *Earth: The Sequel: The Race to Reinvent Energy and Stop Global Warming*. New York: W. W. Norton & Company, 2008.

World Energy Council. "About WEC: What WEC Does." http://www.worldenergy.org/about_wec/135.asp.

Worldwatch Institute. "Making Better Energy Choices." http://www.worldwatch.org/node/808.

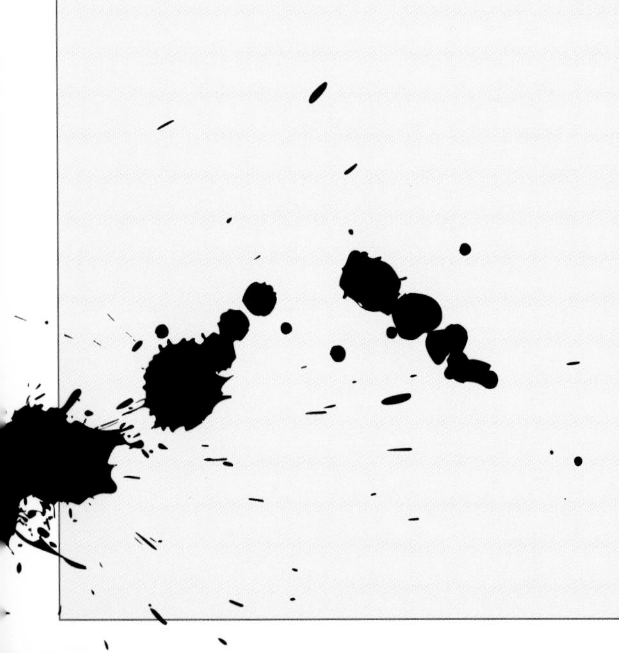

For Further Information

Books

Gorman, Jacqueline Laks. *Fossil Fuels.*
Pleasantville, N.Y.: Gareth Stevens Publishing, 2009.

Kallen, Stuart A. *The Gas Crisis.*
Yankton, S.D.: Erickson Press, 2007.

McLeish, Ewan. *Energy Crisis.*
Mankato, Minn.: Stargazer Books, 2008.

Rodger, Ellen. *Building a Green Community.*
New York: Crabtree Publishing, 2008.

Web Sites

Energy Information Administration: Energy Kids' Page
http://www.eia.doe.gov/kids/index.html

Environmental Literacy Council
http://www.enviroliteracy.org/index.php

Environmental Protection Agency: Climate Change Kids' Site
http://epa.gov/climatechange/kids/difference.html

Energy Star Kids
http://www.energystar.gov/index.cfm?c=kids.kids_index

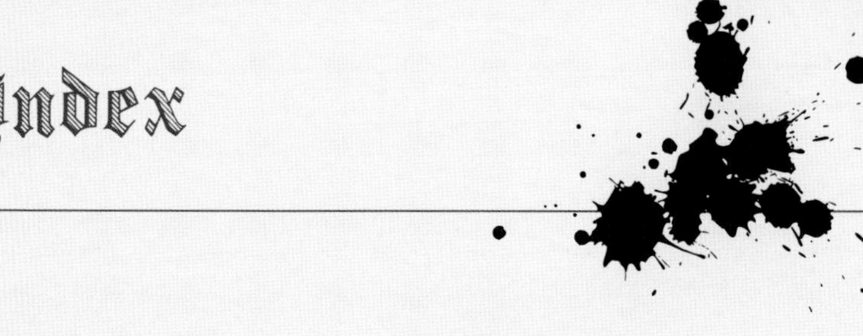

Index